Modern Pig-Keeping

By H. P. Jaques

With Fourteen Illustrations

TENTH EDITION

Cassell and Company Limited
London, Toronto, Melbourne & Sydney

First Edition		-	-	February	1924
Second	,,		-	-	October	1926
Third	,,	(Revised)			October	1930
Fourth	,,		-	-	July	1934
Fifth	,,		-	-	July	1938
Sixth	,,		-	-	September	1940
Seventh	,,		-	-	September	1941
Eighth	,,		-	-	May	1943
Ninth	,,		-	-	August	1945
Tenth	,,		-	-	February	1947

MADE AND PRINTED IN GREAT BRITAIN
AT GREYCAINES (TAYLOR GARNETT
EVANS AND CO. LTD.), WATFORD, HERTS.
1246

Preface

It would be almost impossible for anyone interested in pig-breeding to read the agricultural papers without noticing the amount of space devoted to modern pig-keeping, and especially to the various questions relating to feeding.

Over a year ago it occurred to me that it would be time well spent if I were to collate some of the answers on feeding for my own use, but never with the idea of having them printed. However, several friends and acquaintances having started in this branch of farming, being anxious to have the use of them, together with other notes, and my own views on certain matters in connexion therewith, I was prevailed upon to publish them. Hence their appearance in print and hence the title.

I have to thank the Editors of *The Agricultural Gazette, Farmer and Stockbreeder, Modern Farming*, and *Pigs* for their kindness in allowing me to make use of their respective papers. Also Professor J. B. Wood, C.B.E., M.A., F.R.S., of Cambridge University; Captain Golding, D.S.O., National Institute for Dairying, Reading; Mr. Sanders Spencer, Dr. Rowlands, M.D., Mr. S. F. Edge and Mr. Cuthbert C. Smith, all of whom have readily allowed me to make use of their correspondence. I trust, therefore, the information contained herein will prove of some practical assistance and interest to those who have recently commenced to keep pigs, and perhaps to those who have been longer at the work. In these days of Danish, American and Dutch competition it is of the greatest importance to have pigs ready for the butcher and bacon curer in the shortest possible time, at the least possible cost, of a uniform weight, and of the highest quality. Given pigs of good breeding, the use of correct feeding rations, the open-air system, and catering for the public requirements, this result no doubt can be obtained.

To those requiring a comprehensive book on pigs, I must refer to "The Pig," by Sanders Spencer, for, as already stated, this little work is merely an expansion of my own notes.

The twenty odd years that I have spent farming in Rhodesia and Western Canada, on farms at the "back of beyond," has not been exactly a training ground for book-writing, and this must serve as excuse for any imperfection.

It is only those who have been up against the stock and farming problems in extreme cold and hot climates who can really appreciate the moderate weather of England. It may be fickle, but real long Arctic winters and tropical summers are unknown.

Nature has so blessed these islands that they not only produce the best of all the domesticated animals, but very rarely, if ever, make a complete mockery of the skill and labour of the farmer.

H. P. JAQUES.

Southwold, Suffolk.

NOTE.—Since this book was first published in 1924, importations of hog products have increased very considerably, and are now around £60,000,000 per annum. The embargo placed on *fresh pork* from the Continent means that the farmer has complete protection for this produce. The average value of imported pork for the last few years has been close on £4,000,000. In other words, there is a sure market for a further million pigs.

Experience has brought to light that the chief reason why pigs do not do well, or as well as they should, is the presence of *worms*—the common round worm and the whip worms. It has recently been stated by a great authority that he believes there is not a herd free from these parasites, and the loss they cause to pig breeders and feeders probably amounts to millions of pounds. The *only specific cure* for worms is *Santonin*.

H. P. J.

November, 1926.

Preface to Third Edition

THE writing of the preface for the revised edition of this book naturally gives satisfaction to the author. It also gives him the opportunity of again thanking correspondents from many parts of the world who have written expressing their appreciation of the information contained therein.

To read that in this country, thousands of acres are going out of cultivation, that thousands of skilled stockmen and farm workers are seeking work, that many farms are being run at a loss, and at the same time to realize that over a million sterling is being spent every week for imported pig products does not make pleasant reading.

The amount sent to Denmark in 1929 for bacon only was £27,229,516. There is no doubt that this enormous trade has been built up chiefly by the uniformity of this product. With practically one breed of pig (Landrace—Yorkshire), together with one system of feeding and curing, the Danes have been able for many years to put on the English market a bacon that appeals to the consumer.

In England, on the other hand, there are, as it is generally known, over a dozen breeds of pigs besides numerous crosses. These are fed and fattened according to the fancies of their respective owners. Thus the individuality peculiar to each breed and cross-breed, and the different systems of curing adopted by the many curers, causes the resultant bacon so to vary that the purchaser is never sure of what he will get, so he invariably asks for "Danish," which is always the same in flavour, size, and degree of fatness.

It is therefore obvious that if the English pig-breeders and curers wish to secure this vast trade, they must model their output on the same lines as that of the Danes. Also, be it remembered, that this £27,229,516 for Danish bacon is much less than half the value of this country's annual payments for imported pig products.

In 1929 these imports totalled £60,877,235, or in other words this is an average of £166,787 for every day in the year, or £6,949 for every hour (in 1929 the pig population of England, Scotland and Wales was 2,508,040).

If this trade could be recovered it would mean an average of £1,170,331 for the farmers of each county in England and Wales.

There can be little doubt that this country could raise most, if not all, the pig meat it requires, and secure the necessary food to do so more cheaply than the Danes. This would especially be so if no flour were imported, thereby necessitating the milling here of all imported wheat. Incidentally, this would mean further employment for many thousands of men now on the dole. No wheat offals should be allowed to be exported until the home market requirements have been supplied.

It must be remembered that we are living in an age of mass production and standardization, both of which have for their object, cheapness and quickness. It is the vast majority of the population who buy this imported bacon and not the few to whom the extra cost of English bacon makes no difference.

The standardization of the best cross or breed of pig would of course necessitate the elimination of many breeds (a painful procedure) but either this must be done, or this country must continue to be a gold-mine to the foreign pig-breeder, curer, and importer.

H. P. JAQUES.

Preface to Fourth Edition

THE event of most importance that has of late occurred in the pig world has been the inauguration of The Pig and Bacon Marketing Scheme. Its purpose is to increase the home production of bacon by Government control of imports, the pig producers being paid for their pigs in accordance with the amount realized for bacon. As with a new car, before the scheme can be worked smoothly it requires "running in" and adjusting. If it is a success then there must be a considerable increase of employment on farms, in feed mills, transport, and bacon factories. It may even mean in the course of a few years the evolving of a National Pig—having the valuable assets of being a perfect porker and at a later age a perfect baconer.

With such a pig, and with Government aid to stop dumping, then no longer should it be necessary to spend £50,000,000 a year on foreign products which can be better produced in this country.

The success of the industry depends to a large extent on the manner in which producers support the scheme, both in the matter of making contracts for adequate numbers of pigs and also providing pigs in regularly monthly numbers of the right type and quality.

H. P. J.

Contents

List of Plates

MODERN PIG-KEEPING

CHAPTER I

WHAT'S WRONG?

THAT there is something radically wrong with the pig industry of this country cannot be denied, for how comes it that we had to import during the year 1922 *pig products* to the value of £55,306,027, of which bacon from Denmark accounted for £16,660,616, and *fresh pork* from Holland to the almost unbelievable extent of £2,410,921? Again, why do the provision merchants and grocers prominently label their Danish bacon and hams in the same way that butchers do their prime English and Scotch meat? Is it that they make more profit from the Danish produce, or that they can always rely on the quantity and the quality giving satisfaction and so retain their customers? I think the latter must be the reason, and therefore deduce that the English bacon and hams are of variable quantity and quality. This, then, leads to the question as to whether the fault rests with the breeder, fattener, or bacon curer. I shall answer this question by giving a short account of the Danish bacon industry, and try to prove without disparaging anyone over here that the success of the Danes is largely due to co-operation. Possibly, farmers are tired of having the Danish system of raising bacon and pork held up as a model, but is it not due to the fact that the advice tendered stops when the pig is ready for the curer and not after that stage? It is chiefly by selling practically direct to the retailer that the Danish producer gets most of his profit. Also, by studying our requirements and putting this knowledge into effect, he finds a ready market for his produce.

The English farmer must realize these most important points and so follow the system throughout, if he wishes to participate in a full return for his work and outlay from pig-keeping. Apart from this aggressive foreign competition the fear of swine fever with its accompanying vexatious movement restrictions

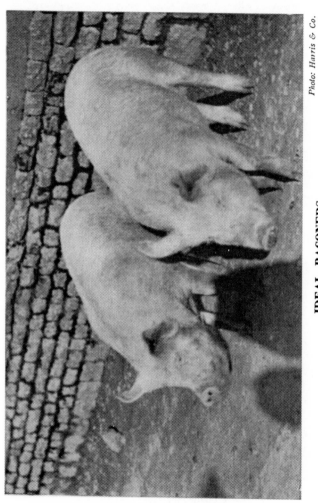

IDEAL BACONERS

(Live weight, 396 lb.—weighed immediately before slaughter)

LARGE WHITE SOW
("Bourne Queen Anne")

LARGE WHITE BOAR
("Worsley Jay")

has undoubtedly curtailed the extension of pig-breeding. During the last ten years the average number of outbreaks has been 2,709; the maximum being in 1914, when 4,356 outbreaks were confirmed by the Ministry of Agriculture. It is satisfactory to know, however, that for the first nine months of 1922 the number was only 908—a very welcome reduction. As the statistics show a steady decrease since 1919, may this not be due to the " open-air system " which has so largely increased during this period?

Another detrimental factor that has played a part in deciding farmers to " ca' canny " in pig-producing has been the violent fluctuations of prices that were so often experienced in pre-war days. To have a large number of fat pigs ready for sale with every possibility of a " copper price " was a most unpopular speculation. Consequently this unstable market must be held as one of the responsible reasons for not keeping pigs on a larger scale.

In 1914 the average price of bacon pigs (first quality) was 7s. 10d. per 14 lb. stone. From then onwards to 1920 it rose to 24s. 8d., whilst for the first nine months of 1922 it averaged 13s. 7d.

These, then, are the more important factors that have paved the way for the Dane and American to exploit our valuable home market.

By taking advantage of co-operation, by the employment of modern methods of production, and by raising the type of pig required by the local bacon factory which knows the public requirements, there is no reason to fear that before long these enormous imports of bacon, pork, etc., will be a relic of the past.

CHAPTER II

THE DANISH SYSTEM

As with us, the Danes at one time sold their pigs through the usual channels, but eventually they realized they were merely working for the benefit of others. This being resented in the early 'eighties, they started a few co-operative bacon factories, a system which now extends practically all through the country.

Certain rules were made; for instance, a customer must be a patron, and guarantee to send his product to the co-operative house, regardless of market and price. The co-operatives pay the transportation, thus the customer who lives at a distance receives just as much net for each pig he delivers as one living near by. Patrons have equal voting rights, elect their officers out of their own rank and file, and at the time of their inauguration none of these, from the president down to the controllers, was salaried. There were neither stocks nor shares sold or given out, but every individual member by signing up placed himself at the back of the scheme, and was fully interested in its success. In this way co-operatives were able to borrow the necessary money for the financing of their schemes.

Almost every town of any size in Denmark has a co-operative bacon factory, but before one can be started it is necessary that the farmers undertake to support it to the extent of 12,000 pigs annually. Owing to the short distance to the slaughter-house it gives the producer the advantage that the hog is slaughtered, if not on the same day as it is shipped, with but little delay.

Every member has a number stamp and an ear-tag, and one or two of these are put on the pig before it is sent to the slaughter-house. After the hog is slaughtered it is weighed and graded first, second or third class, and the customer gets his account according to the day's market price of the grade. Once a year he receives a bonus from the factory's profits, *pro rata* to the number of pigs he has delivered.

The Danish farmers are producing the bacon pig, and much trouble is taken in feeding correct food rations and to get the pigs the right size, the weight limits being 120-160 lb. dressed. Usually a premium is paid for first class, 140-150 lb.; and if overweight a dockage of halfpenny on the whole weight is made for each 10 lb. or fraction thereof overweight. From 170-180 lb. the dockage is a penny, 180-190 lb. three halfpence, and so on. For underweight, however, the line is drawn at every 5 lb.: 120-115 lb. one penny, 115-110 lb. three halfpence.

All the co-operative factories—of which there are nearly fifty—have a central office through which the export goes, and over here they have agents in London and other big centres who sell the product.

That these co-operative bacon factories are successful is proved by the fact that they are steadily increasing whilst those otherwise owned are decreasing.

Thus the producer gets every penny his stuff brings on the market, minus the real handling charges, for by this system the middleman has been almost eliminated.

The great success of the Danish pig industry is more remarkable from the fact that not only must a considerable part of the feed be imported, but also there is the carriage to pay on the finished product to England! Then why is it that the Danish farmer is able to compete so successfully over here, not only with home production but with the rest of the world? The answer is to be found in the fact that he handles and sells his product co-operatively, and thereby obtains every penny his product brings on the market—minus the real handling charges, and in producing a grade of bacon the consumer demands.

Danish bacon comes over " green," and is smoked here by the wholesale provisioners. Moreover, it is delivered carriage paid to the retailer by the wholesaler—a custom that is not followed by the English bacon factories.

In Danish feeding methods all foods are calculated in " food units," with the use of 1 lb. of grain, such as barley, maize, etc., as a basis. In roots and other green feed the food units are estimated on their dry matter, as, for example, 8 lb. mangolds, 4 lb. boiled potatoes, 5 lb. lucerne equal one food unit. In the case of milk 5½-6 lb., and of whey 12 lb. are calculated to have a feeding value of one food unit. In other words, the feeding value of 1 lb. barley, maize meals, etc., has the same feeding value as 5½-6 lb. milk, 8 lb. mangolds, 4 lb. boiled potatoes, or 5 lb. green lucerne. In 1904 Denmark passed a law that all skimmed milk and buttermilk before leaving the creamery must be heated to 176° F., thus destroying any tuberculosis germs.

CHAPTER III

THE OPEN-AIR SYSTEM

THE old practice of keeping pigs in sties, with all the disadvantages as regards health, work and expense, is nearly extinct. The "open-air system," as it is called, now taking its place is not, however, altogether a new idea. In Hampshire, Gloucestershire and in East Anglia especially it has been the custom for generations to run pigs in the open—in fact, Mr. Sanders Spencer practised it sixty years ago in Norfolk. But it certainly is an innovation in many other parts of the country, and where found practicable it has ousted the sty system.

"Pigs reared in the open" is now generally used in the advertisements of breeders, and it shows that the most successful breeders adopt the open-air system as the better method.

Make it a point to visit some of these farms, for there is much to be seen and learnt. The owners are entitled to considerable credit for the part they have taken in recent years in

In introducing a " new breed " into a district uphill work and often disappointment will be encountered. The country is very conservative on this point, although possibly not so impregnable as at one time.

For many years I had to contend against this in Western Canada, where I introduced the Suffolk sheep, and it took many years of hard work, much expense, and the overcoming of many difficulties to make this breed as popular as its rival the Shropshire. It is not an experience I wish to repeat, and I do not recommend others to try it who are beginning pig-keeping.

A sounder and more practical business policy is to take the line of least resistance, or to supply only what is wanted in the immediate neighbourhood. It certainly is for those who cannot afford to wait long before deriving a revenue from their venture.

To those who are about to begin—consider well the following advice : Start with a few gilts or sows of a really good strain, and so gain experience without risking too much capital.

There are many pitfalls that only practical

knowledge prevents one from falling into, and once in, extrication is expensive.

In consequence of the large number of bacon factories now being built in many districts, it will be found, after taking everything into consideration, that the best policy is to learn from them what type of pig they require and to breed accordingly.

As these factories rely on their success in putting on the market a bacon which the consumer demands, they must obtain pigs which will produce this or eventually close down.

To produce such bacon in quantities of a uniform quality, necessitates the greatest attention being given to the right feed.

It would be a wise move on the part of the bacon factories if they were to circularize the farmers as to what they require and the reasons thereof.

Pedigree Breeding.—It is very seductive to read of the high prices obtained for pedigree stock. Many without any knowledge whatever of stock have been tempted to buy expensive pedigree animals, thinking they will thereby be quickly on the high road to fame and big prices from results of their own breeding, but,

BERKSHIRE SOW
("Braishfield Baroness")

BERKSHIRE BOAR
("Highfield Royal Pygmalion")

Photo: Sport & General

ESSEX SOW
("Walden Treasure")

alas! nearly all of them have been doomed to disappointment. Anyone can buy and breed pedigree stock, but it is an art or gift to be able to choose and mate animals whose progeny will make a name for themselves and incidentally their owner. This faculty is possessed by only a comparative few, and the chief reason why some patrons of a breed are so successful is, that they are sufficiently shrewd to employ the services of those who possess this necessary skill. Then, the mere buying and mating of stock is only half the battle in obtaining a herd name, for it must be represented at the shows, there to compete with the pick from already well-known herds.

Showing is expensive, so also is advertising; nevertheless, they are twin necessities. Without the aid of the farm papers it would be extremely difficult for breeders to find a market, or to be conversant with the current affairs of live-stock. Therefore, to become a pedigree breeder of renown requires the greatest skill and patience, and generally a considerable outlay. I say generally, because not all successful breeders to-day were rich when they started, but then they either possessed the aptitude for, or were born and bred to, the work.

In both instances, however, it takes time, often a very long time, and the well-known established herds of to-day have taken many years to attain their position. The Cropwell herd, for instance, has been established for about half a century—and, moreover, these herds are always being improved.

Beginners when starting with pure-breds can always keep one or two of the best of their own raising for the local show, and this may prove to be the first step towards that goal which most pedigree breeders desire to attain, namely, a well-known herd.

Although nearly all such owners have done remarkably well they cannot claim a monopoly of successful pig-breeding, as many have obtained this result from cross-breeding. In fact, the latter far outnumber the former, and no doubt do obtain in most instances with certain breeds the combined merits of the two. But it must be borne in mind that this result is mostly due to the perfection to which the pedigree breeders have brought their respective herds.

bowels open, which is also accelerated by a little exercise.

In due course gradually get her on to full feed, which should take from ten to fourteen days. If this precaution be not taken and heavy feeding be given earlier, illness of the little pigs is almost bound to occur.

A constant source of annoyance to the sow is from two dark teeth on each side of both the upper and lower jaws of the little pigs, which they often possess at the time of their birth. These invariably make the teats sore and frequently have a dire result; it is therefore better to take the precaution of examining the mouths as soon as possible after birth and to clip off such teeth.

Weaning

In this matter also there is a diversity of opinion. Some breeders do not wean the pigs until they are three months old, whilst others insist on this being done at eight weeks. In Denmark pigs are weaned at four weeks.

As good results are obtained from doing

so at both ages, it cannot be argued that one is right and the other wrong. At the same time I believe that the majority of breeders favour the longer period, as for one thing by the time pigs are three months old they have almost been naturally weaned, and are quite accustomed to trough feeding, whereas those a month younger are not so able to fend for themselves. Consequently they will require extra attention, which means more labour and expense, which items all are striving to reduce. Also there is no fear of udder trouble, as there may be from weaning at a time when sows should be giving plenty of milk.

If a sow is bred the first time she comes into season, after her pigs have been weaned she will practically always hold to that service. A sow comes into service about every twenty-one days.

lucerne, feed peas. There is no necessity to soak peas, as even very young pigs can masticate and digest them without their being softened.

Pollards.—Ratio, 1 to 5. Similar to middlings, but more fibre and ash. Not advisable to give to young pigs.

Rice Meal.—Ratio, 1 to 9. High in oil, but 50 per cent. of the oil is generally not digestible, being of a fatty acid nature. Produces poor quality meat.

Roots.—Artichokes are a good pig food, they do not require to be cooked. Can be grown in any odd piece of waste land for pigs to grub out. Mangolds and swedes are at times beneficial if used in moderation, the former during and after March, and the latter from December to March.

Sharps.—Ratio, 1 to 4. Best for sows after farrowing for two or three days. It is not good for making pork.

Wheat.—Ratio, 1 to 7. Should not form more than 25 per cent. of total ration. It is a very rich food. If finely ground it forms a sticky paste when wet, and is therefore better fed when rolled or crushed very coarsely.

READY RECKONER FOR COST OF FOOD

Per ton.						Per lb.
£	s.	d.				d.
2	6	8	...	...	...	¼
4	13	4	...	...	...	½
7	0	0	...	...	...	¾
9	6	8	...	...	...	1
11	13	4	...	...	...	1 ¼
14	0	0	...	...	...	1 ½

The above reckoner saves considerable calculation in determining the cost of a ration. With this information and knowing the increased weight it is easy to find the cost per lb. of feeding for pork or bacon productions; providing, of course, that particulars are kept of the quantity fed.

To know this cost is a factor in making pig-keeping a success.

CHAPTER VIII

FEEDING

No very hard and fast rule can be laid down as to the exact amount of feeding a pig should have daily. Considerable judgment must be exercised, but as a guide the following amounts are about correct :

					Albuminoid Ratio.
Weaners, ½ to 2 lb. per day ...			...	...	1 : 4
Four months,	3 ,,	,,	...	...	1 : 5
Five months,	4 ,,	,,	...	...	1 : 5½
Six months,	5 ,,	,,	...	...	1 : 6
Seven months,	6 ,,	,,	...	...	1 : 6

Gilts should never have more than 6 lb per day until they have farrowed.

Sows suckling young until they are twelve weeks old up to 10-12 lb. per day.

Boars, allow 1 lb. of feed for each month of age up to 10 lb. for ten months of age.

Fattening pigs are generally fed three times daily, but towards the end of this period give as much food as they will clean up.

If soaked feed is given in excess and remains in the trough, it becomes sour and is responsible for many intestinal disorders.

As a rule, it takes nearly 4 lb. meal to obtain 1 lb. increase of live weight—i.e. from weaning up to around 200 lb.

By weighing the pig once a week or every ten day proof can be obtained as to whether the ration given is satisfactory or otherwise.

With the exception of feeding pigs on soaked maize, peas and beans when on grass, or running them over the stubble, the feeding of whole grain is wholesale waste. That this is so can easily be verified by looking at the excreta, when it will be seen that quite a considerable percentage has passed through whole.

Grain food, therefore, should be ground as finely as possible. This not only enables the pig to extract more easily all the nourishment possible, but saves much energy on the part of the organs, which must have extra "fuel" to do the work of digesting whole and coarsely ground meal.

Those keeping pigs on a considerable scale and who do not possess the necessary machinery, will find it sound economy to invest in a small oil engine and a grinder to do the part that the molars should do, but don't.

Another advantage is, that one knows

exactly the quality of the food, which is not always the case with bought meal. It also saves considerable time and labour of carting to and from the mill, which is often some distance away.

Until recently it has been the custom to soak the meal rations for twelve hours preparatory to feeding the pigs with them, generally twice a day at stated times.

It is important that feeding should always be punctual to time. The non-observance of this rule is often one of the factors in those cases where through the neglect of details the owner finds he is keeping the pigs and not they " keeping him."

Now, however, dry-feeding in many parts of the country is being given a trial, and it will be interesting and instructive to know the general result of this form of feeding, both from the feeders' and bacon-curers' point of view, as against the long established wet mash.

The great drawback to feeding a wet mash is experienced during cold weather, when after soaking it is almost ice cold, thus giving much extra work to the system in bringing the food up to blood heat before it can be digested.

A further disadvantage of this mash is that the pigs, when fed, make a great rush and fierce competition takes place to secure advantageous positions. Consequently the stronger obtain more than their fair share, and all consume far more liquid than they require.

In dry feeding, therefore, ice-cold feed is obviated, crowding and bolting are negligible, and only sufficient water is taken at the time to assist mastication.

By placing the water-trough (a plentiful supply of water is absolutely indispensable with dry feeding) at some distance from the feed, weaker pigs are able to get their share whilst the others are away at the water-trough.

One drawback, however, to the use of troughs for dry meal is that the wind will often blow it out, and another that the pigs can more easily scatter the meal.

The use of a good self-feeder prevents this waste, and has proved successful for fatteners and perhaps for young breeding stock. As feeders are made to hold sufficient food for several days the labour of feeding has been reduced to a minimum. It is necessary, however, to inspect the feeder once or twice a day, as the meal may clog in the hopper, and so cease to drop into the trough.

In the case of older gilts, sows after farrowing, and boars, if the feeder be kept open all the time, they can, of course, obtain an unlimited supply of meal. This is not only quite unnecessary but is ruinous, and further, most of the animals under this condition quit grazing and so lose the exercise which it affords, thus defeating two important objects of the open-air system.

Judging by the correspondence appearing in the agricultural Press, nearly every breeder using a feeder has experienced this trouble. To surmount the difficulty they either shut off the supply during certain times, or put into the hopper only a day's requirements, keeping the water as far away as possible.

The necessity of having to manipulate the feeder in this way somewhat detracts from its value as a labour-saver. As certain classes of pigs with free access to food at all times will eat more than is economical for the gain they make, it seems that a self-feeder which will prove suitable for *all* classes will probably never be evolved.

CHAPTER IX

IF only the good individuals of a pure-breed were registered, there would not be the pure-bred " scrub " so often seen at sales and elsewhere. A great mistake is to presume that because the sire and dam are registered pedigree stock, all their progeny should likewise be recorded. There is considerable variation with pure-bred pigs, and it is incumbent on the breeder's part to register only such as possess outstanding merit. To do the contrary not only damages the breed, but to the breeder it does more, for sooner or later the herd loses its reputation, and with that his sales gradually cease to exist.

There is considerable competition in the pure-bred market, and only those who systematically sell really good individuals of a breed can continue to " carry on " successfully.

It must be obvious that without some system of marking it would be quite impos-

44

WESSEX SADDLEBACK SOW

TAMWORTH BOAR
("Knowle Sunstar 2nd")

GLOUCESTER OLD SPOTS SOW
("Sonderna Mascot")

Photo: Sport & General

MIDDLE WHITE BOAR
("Wharfedale Deliverance")

CHAPTER XI

THE ABSORPTION OF FOOD

BEFORE proceeding to give some rations suitable for the various ages of pigs, it will not be out of place to quote from one of Dr. Rowlands' *Agricultural Gazette* lectures on the manner in which the pig absorbs the food into its system :

"*In the mouth,* the carbohydrates (or starch) are affected by the saliva.

"*In the stomach* starch is not affected, but proteins are by means of the pepsins and the hydrochloric acid.

"After being digested in the stomach food passes on to the small intestine, and there the bile joins with it. When starch and this digested protein come out of the stomach, the protein is almost finished for good, but the starch begins to be digested in the *small intestine.*

"After leaving the small intestine it passes

into the large gut, the duty of which is chiefly to absorb the nourishment and the liquids from the food, so that the excreta become nearly solid.

" If pigs are fed on more protein than can be digested, trouble sets up. If food passes hurriedly through the system, and protein goes out of the stomach undigested into the intestine, it ferments, thus causing ' pot-bellied ' pigs.

" A ratio of one part of protein to four of starch or carbohydrates is almost as much as an animal can digest.

" Recapitulation

" *In the mouth* the starch is partly digested.

" *In the stomach* only the protein and albumen.

" *In the intestine* itself entirely starch, except that there the bile mixes with the fat and forms a soap, and as soap the fat is absorbed into the system.

" The chief use of the bile is to digest the fat. Albuminoids or proteins are materials which contain nitrogen, of which fish meal is a good example."

Lately there has been some correspondence in the farming papers, owing to complaints arising from the fishy flavour of bacon, caused through the feeding of fish meal. As some breeders report they have used it right up to the last meal given to the pig before slaughter without receiving any complaint, it probably arises from using a meal not made from *white* fish, or from feeding far too much.

However, if fish meal is an objection, dried blood and meat meal can be used, but in lower proportions, as it is richer in albuminoids, as will be seen by the following comparison :

		Fish Meal.		*Blood Meal.*		*Meat Meal.*
Crude Protein	...	50.0	...	72.7	...	67.2
Pure ,,	...	46.0	...	63.6	...	63.6
Oil ,,	...	4.2	...	0.8	...	12.5

CHAPTER XII

IN modern pig-keeping the food rations are fed according to their correct nutritive ratio—i.e. the proportion of proteins or repair substances to carbohydrates and fats. Formerly it was done by mere guesswork, to-day with complete accuracy. A ratio of 1 part protein to 4½ of carbohydrates or starch is about as much as it is advisable to use, although for short periods most excellent results have been obtained from a mixture of 1 part to 2·8.

The composition and nutritive value of feeding stuffs is to be found in " Rations for Live Stock," by Professor J. B. Wood, obtainable from the Ministry of Agriculture. It is from this source of information that the ratio of any mixture can be determined. At the time when Dr. Rowlands was giving his lectures at Keston, Kent, a controversy arose over his method of computing the ratio in a mixed ration.

As no two correspondents in the *Agricultural Gazette* appeared to agree as to the correct method, I wrote to Professor Wood, and he kindly sent me the following letter: " I am afraid there is no short formula for calculating the nutritive ratio of a mixed ration. The following example, however, will show how the calculation is made.

"Barley meal contains on the average 6.5 per cent. digestible protein, 1.2 per cent. digestible oil, 62.2 per cent. digestible carbohydrates, and 2.5 per cent. digestible fibre. The proportions in which the three foods are mixed in the compound you mention in your letter are: 7 parts barley meal, 2 parts fish meal, and 1 part dried yeast. Multiplying the percentages in barley meal by seven, the quantities work out as 45.5 parts digestible protein, 8.4 digestible fat, 435.4 digestible carbohydrates, and 17.5 digestible fibre. Taking the composition of fish meal as given in ' Rations for Live Stock,' and multiplying these by two, the quantities work out at 100 parts digestible crude protein and 8.4 parts digestible fat, and there is no carbohydrate or fibre. Dried yeast contains 41.6 digestible protein, 0·2 digestible oil, and 29·2 digestible carbohydrate, and there is no fibre. Since only 1 part of dried yeast is used, there is no need for multiplication.

" The quantities in 10 parts of the mixture add up as follows :

	Digestible Protein.	Digestible Fat.	Digestible Carbo- hydrate.	Digestible Fibre.
7 Parts Barley ...	45·5	8.4	435·4	17·5
2 Parts Fish Meal...	100	8.4	0	0
1 Part Dried Yeast	41.6	0.2	29.2	0
	187.1	17	464.6	17·5

Multiply the fat by 2.3 to reduce it to carbohydrate equivalent = 39.1.

" The total carbohydrate equivalent is therefore 464.6 + 39.1 + 17.5 = 521.2, and the ratio is 521.2 + 187.1 = 1 : 2.8 almost exactly."

Taking the above as an example, no one should have any difficulty in working out the A.R. of any ration, and although it necessitates a somewhat considerable figuring, it will be correct.

Besides these concentrated foods, potatoes, especially " chats," are often fed to pigs, and this is the only feed, so far as I have been able to discern, that requires to be cooked. Potatoes, together with barley meal and a little maize gluten, which should be scalded, make splendid bacon, but they are not very good for sows with pigs, or young pigs until twelve weeks old. From four to eight pounds per day suffices, but if the latter is exceeded then the maize or barley meal must be reduced.

6 lb. of potatoes equal 1 lb. of meal.

Other roots given raw are a great assistance in the feeding of sows, but mangolds are not suitable from the beginning of October to the end of March. In making up a food ration the following materials must be supplied :

1. Proteins or albuminoids.

2. Carbohydrates or starches.

3. Certain oils or fat.

4. Mineral salts of some kind, especially phosphate of lime. Calcium phosphate, which is in basic-slag, is taken up into the plant and thence into the animal.

A narrow ration is one which contains a relatively high percentage of protein furnished by such feeds as fish and meat meals, dried yeast, etc.

A wide ration is one rich in carbohydrates and therefore limited in protein. Barley and maize meals are examples of carbohydrate feeds.

Narrow rations are conducive to rapid growth, and are generally fed to young animals.

Wide rations are fed to fattening animals.

A balanced ration may furnish all the feed nutrients, yet the system of the pig craves mineral matter. In order to make sure that the pig has an abundant supply, free access should be given to a Mineral Mixture. One very popular with hog-raisers in the States and recommended by their Department of Agriculture is :

Charcoal, 1 bushel.

Hardwood Ashes, 1 bushel.

Salt, 8 lb.

Air-slaked Lime, 4 lb.

Sulphur, 4 lb.

Pulverized Copperas, 2 lb.

Mix the lime, salt and sulphur thoroughly, and then mix with the charcoal and ashes. Dissolve the copperas in one quart of hot water and sprinkle the solution over the whole mass, mixing it thoroughly. Keep some of this mixture in a box before the pigs at all times, or place in self-feeder.

As a bone strengthener, and an aid to lessening greatly the desire to "root," a mineral ration cannot be too highly recommended.

CHAPTER XIII

THESE, and the subsequent examples of rations, are only some of the suitable combinations that can be given profitably to pigs. They have all been recommended by those with a long experience in feeding, and therefore may be used with every assurance of obtaining good results. Also, they give a good idea of the approximate quantities of the different foods used to suit the various stages and purposes of a pig's life. The albuminoid ratios have been worked out from the tables in Professor Wood's book, " Rations for Live Stock," and from his method of calculating mixed rations as given in the preceding chapter.

60% Maize Meal
25% Bean Meal
10% Middlings
5% Fish Meal   A.R. 1 : 4.9

55% Maize or Barley Meal
25% Bean Meal
15% Palm Kernel Meal
5% Fish Meal   A.R. 1 : 4.5

70% Maize Meal
15% Palm Kernel Meal
10% Bean Meal
5% Fish Meal   A.R. 1 : 5.6

75% Maize Meal
20% Bean Meal
5% Fish Meal   A.R. 1 : 5.7

40% Maize Meal
20% Barley Meal
25% Bean Meal
10% Middlings
5% Fish Meal   A.R. 1 : 4.9

For In-pig Gilts :
65% Maize Meal
25% Sharps
10% Fish Meal   A.R. 1 : 5

65% Maize Meal
20% Palm Kernel Meal
10% Sharps
5% Fish Meal   A.R. 1 : 5.6

The last two are for in-pig gilts at grass
due to farrow in about seven weeks. Barley

meal can be used in lieu of the maize meal. Use about 5-6 lb. per day or less if grass is fair.

Note.—Equal parts of maize and beans thoroughly soaked—if possible, germinated—can be used for in-pig sows and gilts at grass at any time, about 4 lb. per day.

If peas are used in place of beans, use two parts of peas to one of maize. In-pig sows at grass do not require sharps, but a week or so before farrowing omit bean meal and palm kernel meal and substitute sharps.

A week before gilts are due to farrow it is as well to feed mainly on sharps and vegetable food, also for a few days after farrowing, then a mixture of

    25% Sharps
    35% Maize Germ Meal
    30% Barley Meal
    10% Fish Meal   A.R. 1 : 4.5

which can be continued until the pigs are three months old.

The A.R. shows the relative proportion of albuminoid, protein or flesh formers to the carbohydrates, such as sugar, starch and fibre, fats or oil which supply the fuel only.

Thus A.R. 1 : 4.5 means one part albu minoid (flesh former) to 4.5 parts carbohydrates (fuel substances).

The proportion given varies according to the age of the pig (*see* page 39), and for what purpose it is being fed.

Unless the food is mixed in approximately correct proportions, it is practically impossible in these days to feed pigs economically and therefore profitably.

CHAPTER XIV

RATIONS FOR SOWS WITH YOUNG

45% Sharps
45% Maize Germ Meal
10% Fish Meal   A.R. 1 : 4

40% Palm Kernel Meal
20% Barley Meal
20% Middlings
10% Maize Germ Meal
10% Fish Meal   A.R. 1 : 3.4

20% Barley Meal
40% Maize Meal
20% Bean Meal
15% Middlings
 5% Fish Meal   A.R. 1 : 5

25% Barley Meal
40% Maize Meal or Maize Germ
25% Palm Kernel Meal
10% Fish Meal   A.R. 1 : 4.5

30% Sharps
35% Maize Germ Meal
25% Barley Meal
10% Fish Meal   A.R. 1 : 4.3

Note.—When sows have farrowed, sharps are best for two or three days, and then half sharps and half maize germ meal with eight ounces of fish meal. When pigs are a month old, increase proportion of maize germ or barley meal to take the place of half the sharps.

Potatoes should not be given either to sows with pigs or to the young pigs until quite twelve weeks old. Potatoes should always be cooked.

To stimulate flow of milk from the sow the following ration has been recommended:

   30% Cotton Seed Meal[1]
   50% Barley Meal
   10% Pea Meal
   10% Fish Meal   A.R. 1 : 2.9

[1] In Mr. Tod's opinion a cotton seed meal ration is too high in albuminoids.

CHAPTER XV

RATIONS FOR YOUNG PIGS

20% Sharps
30% Barley Meal
40% Maize Germ Meal
10% Fish Meal   A.R. 1 : 4.6

40% Middlings
40% Maize Germ Meal
20% Barley Meal
2-3 oz. Blood (dried)   A.R. 1 : 6

The above two for newly-weaned pigs.

33½ Barley Meal
33½ Maize Meal
33½ Pea Meal   A.R. 1 : 6

Above ration is the Suffolk Standard mixture.

65% Barley Meal
25% Sharps or Middlings
10% Fish Meal   A.R. 1 : 4.7

This is a splendid mixture and is recommended for pigs of all ages. As pigs get

Photo: E. David

OLD GLAMORGAN SOW

Photo: Sport & General

LARGE BLACK BOAR
("Carnewood Someday")

LINCOLNSHIRE CURLY-COATED BOAR
("Sleaford Chum II")

CUMBERLAND SOW
("Agnes")

older, maize meal can be substituted for barley if cheaper.

Probably this is the most popular of all the mixtures.

50% Barley Meal
25% Sharps
15% Pea Meal
10% Fish or Meat   A.R. 1 : 3.9

Little pigs take to this mixture readily, and will fatten so well that if desired they can be sold as porkers at twelve weeks old. As the age of the pig increases, the husk need no longer be removed from the barley meal, the sharps may be left out of the ration, and if available, bean meal may be substituted for pea meal. This should be fed at three weeks, when they may be allowed as much as they can clean up.

Note.—Feed young pigs four times per day, and, if possible, give the last meal at 9 P.M.

CHAPTER XVI

20% Wheat Meal
50% Maize Germ Meal
20% Sharps
10% Fish Meal   A.R. 1 : 3.5

40% Barley Meal
30% Wheat Meal
20% Bean Meal
10% Fish Meal   A.R. 1 : 3.9

40% Maize Meal
35% Bean Meal
20% Palm Nut Cake
5% Fish or Meat Meal   A.R. 1 : 3.8

20% Sharps
45% Maize Germ Meal
25% Barley Meal
10% Fish Meal   A.R. 1 : 4.6

Above ration for three months old pigs
intended for showing. Gradually increase

the barley meal at the expense of sharps and
maize germ.

50% Barley Meal
20% Wheat Meal
20% Bean Meal
10% Fish Meal   A.R. 1 : 4

70% Barley Meal
20% Fish Meal
10% Yeast Meal   A.R. 1 : 2.8
                         (See page 55)

This is Dr. Rowland's ration which he
stated gave 31 lb. increase for 51 lb. of
food. It led to considerable controversy.

The Suffolk Standard mixture, *see* page 66.

70% Barley Meal
25% Pea or Bean Meal
5% Fish or Meat Meal   A.R. 1 : 5.1

A very successful mixture.

For Pigs on Rape:

65% Barley Meal
25% Palm Kernel Cake
10% Fish Meal   A.R. 1 : 4.2

For Gilts on Grass:

35% Barley Meal
35% Wheat Meal
25% Bean Meal
5% Fish Meal   A.R. 1 : 4.5

CHAPTER XVII

50% Sharps
25% Maize Meal
15% Barley Meal
10% Fish Meal   A.R. 1 : 4.1

40% Barley Meal
30% Maize Meal
20% Palm Kernel Meal
10% Fish Meal   A.R. 1 : 4.7

40% Maize Meal
30% Barley Meal
15% Middlings
15% Bean Meal   A.R. 1 : 6.8

40% Maize Meal
30% Barley Meal
20% Bean Meal
10% Malt Culms   A.R. 1 : 6

35% Barley Meal
30% Maize Germ Meal
15% Bean Meal
20% Middlings   A.R. 1 : 6

Above ration for pigs six months old.

40% Barley Meal
35% Maize Meal
20% Palm Kernel Cake
5% Fish Meal   A.R. 1 : 6.2

Above ration for pigs forward in flesh.

30% Barley Meal
30% Maize Meal
30% Sharps
10% Fish Meal   A.R. 1 : 4.7

Above is a good general ration and is fairly suitable for all purposes, but for young pigs use extra sharps in place of maize.

70% Barley Meal
20% Pea or Bean Meal
10% Fish Meal   A.R. 1 : 4.4

Above is one of the best and simplest fattening rations. This can be gradually worked up to from the previous ration, so that the pigs are receiving it during the last month.

35% Maize Meal
30% Barley Meal
20% Bean Meal
15% Middlings   A.R. 1 : 6

65% Maize Meal
20% Bean Meal
15% Middlings   A.R. 1 : 6.4

Note.—An excessive proportion of potatoes, maize or sharps in the ration results in bacon being soft and oily.

An excess of beans or bean meal causes the lean portion of the bacon to be hard.

Palm kernel cake tends to harden the fat, and also makes pigs weigh well when slaughtered. It is a much cheaper food for fattening pigs to the extent of 20 per cent. of their concentrated food than coconut cake.

Pigs intended for pork should be fattened and ready at four months of age. Obtain local requirements as to weight. Pigs intended for bacon should weigh about 230 lb., and be brought to that weight in eight months. They must be long and wide in the back, deep-sided, have thick hams and be light in the shoulder; these being respectively the expensive and cheaper parts. The ham cannot be too meaty, and a high percentage of lean meat is required.

Approximately there is a loss of 25 per cent. between live and dressed weights.

ROTATION OF FORAGE CROPS ENABLING PIGS TO BE FOLDED ALL THE YEAR ROUND
(MR. STANLEY WILKIN)

To be sown in	Seed per acre	Variety	Thin out to	Ready for grazing	Will feed 60 six months old pigs
					DAYS
Feb. or Mar. Mar. to July	8 lb. 10 ,,	Marrow Stem Kale Rape (broadcasted)	30 in. —	Sept. to Christmas Ten weeks after sowing	120 60
March April	4 ,, 8 ,,	Kohl Rabi Thousand Headed Kale	20 in. 24 × 24 in.	Oct. to Mar. Jan to April	90 100
July August	4 ,, 2 bushels	Coleworts Tares and Rye (equal parts)	18 in. —	Mar. to Sept. April	40 60
September	1 bushel	Each Tares and Beans and Rye or Oats	—	May	60
October	1 bushel	Maple Peas and Oats	—	May and June	60
November	1 bushel	Each Tares and Beans and Oats or Wheat	—	May and June	60

CHAPTER XVIII

WEIGHING

To check the weight of bought feeding stuffs, etc., to weigh the food rations, to know exactly from week to week the result therefrom by weighing the pigs and to be on the right side when selling, a weighing machine is indispensable.

The weighing of bought food prevents paying for short weights; the weighing of food rations prevents, to a large extent, over- or under-feeding; the weighing of pigs tells if a profitable food ration is being given, whilst to know their exact weight is better than the usual method of estimating. Without weighing, selling of pigs for slaughter is a guessing competition. After two years' experience as a Live Stock Sub-Commissioner, which work included investigating the results of graders estimating the dead weight of sheep, as it was under the Live Stock Control, I have no hesitation in saying farmers would gain considerably if the weighing of stock were the rule rather than an exception. As butchers always sell their meat over the scales, why should not farmers sell their live "meat" in a similar manner?

I had scores of instances during the Control of comparing the estimated dead weight with the actual results, and more often than not it was in favour of the butcher. Evidently they make a point of *under* estimating when buying—and it must be exceptional when they do the contrary. Both in Canada and the States all slaughter stock is sold over the scales, the price quoted for the different grades or quality means so much per lb. or 100 lb. live weight. By this simple system it can be seen at a glance what live stock is realizing at the different markets throughout the whole country, but the chief consideration is that the farmer receives the exact worth of his stock according to the market price.

CHAPTER XIX

DISEASES

APART from swine fever and foot-and-mouth disease—which, should an outbreak occur, must be reported to the police—tuberculosis is, I believe, more prevalent among pigs than is generally supposed. In the United States about 10 per cent. are affected in some degree, but owing to the custom of private slaughter-houses in England, it is quite impossible to obtain figures that are even approximately correct. The most probable causes of infection are from tubercular cattle, milk, garbage, slaughter-house refuse and sunless damp quarters. The best means of prevention, then, is to keep pigs clear of cattle, unless they have passed the tuberculin test, to quit feeding on such material, and to adopt, if possible, the open-air system. After many years' experience of raising hogs in Western Canada, I think pigs, if away from sties, keep, as a rule, wonderfully free from disease. I never came across any of the serious diseases, but, of course, have had experience of a few of the lesser ailments, the result, no doubt, of faulty feeding. Thanks to one of those practical circulars issued by the Dominion Experimental Farms, which I found so useful in overcoming the lesser complaints, I do not think I can do better—permission having been readily given—than to quote from same.

Constipation
Particularly to be guarded against with pregnant and milking sows.

Cause.—Lack of exercise and of succulent food.

Treatment.—Remove cause. Give 2-4 oz. raw linseed oil daily in slop for mature animals. If no effect, give as drench 4 oz. Epsom salts. Avoid drastic purgatives with the milking sow.

Mr. Tod's opinion is that the easiest remedy for constipation is greenstuff or bran.

Diarrhœa (*Scours*)
Common and fatal with young pigs particularly.

Cause.—Over-feeding the sow after farrowing with rich foods. Sudden changes in feed.

Use of decomposed or sour slop. Nervousness and irritability in the sow.

Treatment.—Change feed. Give 15-20 grains iron sulphate to the sow in slop night and morning. Mix lime-water with slop or supply where sow can reach it a mixture of iron sulphate, sulphur and salt (equal parts) with four times quantity of charcoal. For young pigs give 2 oz. castor oil.

Indigestion

Symptoms.—Unthriftiness, poor feeding, arched back.

Cause.—Over-feeding; feeding decomposed slop or swill containing injurious substances.

Treatment.—Remove all food for twelve hours, give 4 oz. castor oil, feed lightly on bran and shorts with green food or roots.

Thumps

Symptoms.—Usually seen in young pigs, dullness, constipation or diarrhœa, short breathing with a peculiar thumping noise.

Cause.—Disordered digestion due to too much concentrates in ration or too much feed in combination with lack of exercise.

Treatment.—Preventive largely. Provide exercise, forcing it where necessary in cases of heavy milking sows, by removing pigs to another pen for an hour or so daily. Restrict food of sow. Apply these measures at first sign of over-fatness or sluggishness and thumps will not appear. With weaned pigs reduce concentrates, increase skim milk and force exercise. In individual cases castor or linseed oil. Difficult to treat.

Crippling

Symptoms.—Often confounded with rheumatism, stiffness and lameness, generally of hind legs. Animal lies most of time until walking becomes impossible. Appetite disappears and death ensues.

Cause.—Strong food and too much of it, lack of exercise, damp quarters due to bad ventilation, wet floors, filth. Usually a combination of all.

Treatment.—Prevent by supplying right conditions. Exercise outdoors. Feed as already outlined. If condition is advanced force exercise, give 2-4 oz. Epsom salts in pint of water, repeated in twenty-four hours. Feed, in small quantities, milk, bran and shorts with roots or green feed. Give two tablespoonfuls daily, of sulphur, Epsom salts and charcoal, equal parts.

Rheumatism

Symptoms.—Lameness, stiffness, pain and swelling in joints.

Cause.—Almost invariably due to damp quarters, due to wet floors, filth, or damp walls and impure air, the result of poor ventilation. Heavy feeding, in conjunction, complicates matters.

Treatment.—Difficult in advanced cases; see treatment for "crippling." Give salicylate of soda three times daily in feed, 20-30 grains to the dose. Use liniments or blistering ointments on affected joints. Give dry quarters and plenty of bedding. Prevent, by adopting outdoor methods for all but fattening and very young stock.

Inflammation of the Udder

Treatment.—Milk two or three times daily. Give small dose of Epsom salts and feed on sloppy diet. Apply ointment as follows, kneading well. Extract of belladonna, gum camphor, 1 dram each, vaseline 3 oz. Apply hot fomentations.

Parasites (*Internal*)

For intestinal worms, give turpentine, 1 teaspoonful for every 100 lb., in raw linseed oil, as a drench, after having removed all food for at least twelve hours, or administer in slop. Follow by physic of Epsom salts. Prevent by allowing pigs access to mixture of charcoal, wood ashes and salt.

Lice

Apply crude castor oil, crude petroleum, a mixture of raw linseed oil 2 parts, kerosene $\frac{1}{2}$ part, or fish oil 12 parts, creolin or paraffin oil 1 part. Disinfect and clean quarters if infested.

To Drench a Pig

Use care. Go slowly; back it into a corner, raising the head slightly. Attach a piece of hose, six inches long, to a small, long-necked bottle. Insert hose into pig's mouth and pour contents slowly. The pig chews the hose, receiving the dose naturally and lessening the danger of choking.

CHAPTER XX

BEFORE turning pigs into a pasture or field be sure the fences are pig proof.

The corner posts practically make a fence; if loose, fence is slack, and therefore useless for " holding " pigs.

A bottom strand of barbed wire prevents pigs lifting the woven wire fence.

Paddocks of half an acre in extent are about the right size for farrowing sows, and where they are kept until removed at weaning.

Rest the paddocks occasionally—slag when necessary.

Ideal paddocks are those containing a good growth of wild clover, facing the south, and protected from the north by a belt of trees, or by the lay of the land.

Huts should be approximately 6 by 8 feet, with floor raised 6 to 8 inches above ground. Doors not less than 26 inches wide; narrow doorways are an abomination.

Keep the floor dry and well littered with wheat straw. It is preferable if huts are so placed that they face the south-south-east.

Frequently move the feeding trough as it saves a patch of grass from being destroyed, and causes the manure to be better distributed.

"The boar is half the herd." He represents 50 per cent. of the breeding stock, and therefore is the most important individual in the whole herd. See to it that he is a better representative *of the breed* than any of the sows.

Select a boar at least eight months old. It is very difficult to foretell the future outcome of a weaner.

Always choose a boar that stands well up on his toes, a sure indication of having *strength and bone*. Beware of weak hind quarters.

Sows are approximately 115 days in-pig.

Oestrums generally occur at intervals of three weeks, commencing about five days after the pigs have been weaned. Do not miss this service.

As to the age when gilts should be mated even well-known breeders are at variance, some having them mated at eight months, whilst others prefer to wait until they are a year old.

Best time to mate gilts is during October and the early part of January, then both litters come at a good time. It is important to breed from sows having not fewer than twelve teats, with the front ones well forward.

For breeding stock and stores pasture provides for growth of bone, and muscle, and general vigorous health, and affords considerably cheaper gains.

Pigs farrowed later than July are rarely such good doers as those farrowed earlier.

Never interfere with or disturb a sow farrowing, unless absolutely necessary, and then only with great care. Remove and destroy the after-birth.

After farrowing see that the sow can get a drink of water not too cold. She hardly requires any food for several hours, and then only a thin warm slop of bran and middlings. Do not feed within two hours of farrowing.

A week at least should be taken before putting the sow on full ration.

The small pigs are a good guide as to whether she is being fed properly.

If the sow be over-fed resulting in a heavy flow of milk, the little pigs will nearly always scour.

If under-fed—not making enough milk—the pigs will follow her, and be continually pulling at her teats. Govern the feed accordingly.

Grease the tails—especially in cold weather —of newly-born pigs : this prevents the end from sloughing off, an incident very apt to occur without this precaution.

Should the sow be worried by her pigs suckling, cut or break off their teeth, which are often dark in colour.

The best time to castrate pigs is between six and eight weeks of age—i.e. before they are weaned.

It should never be done on a rainy day. A disinfectant should always be used.

Never wean under eight weeks; many of the most successful breeders keep the pigs with the sows for three months.

To check milk secretion at weaning, feed solely on dry oats, giving plenty of clean water to drink.

It is more profitable for a sow to raise eight

to ten good pigs than to bring up more and not do them as well.

Feed fattening pigs three times daily—no more than they will clear up. Use judgment in feeding.

It is a good plan to have pigs fed their final meal at the last thing at night, thus causing them to empty their bowels and bladder—especially is this so during the winter months.

Changes in food should be made gradually.

A tablespoonful per head per day of cod liver oil to backward pigs is very valuable, supplying the necessary vitamins.

Large and small pigs should not be fed together.

Whether the feed be given soaked or otherwise, a supply of clean drinking water is absolutely essential.

Many pure-bred pigs with faults that are not desirable to reproduce should be fattened and consigned to the butcher. It is a mistake not to do so, for not only the herd but its reputation will soon suffer in consequence.

The use of expensive condiments cannot be recommended, the best is that supplied by

Nature—the sun, fresh air, natural herbage, and exercise.

Separated milk when fed alone is not suitable, as it is deficient in carbohydrates and fats. Maize meal will make up this deficiency, using about 3 lb. to each gallon of milk. This quantity is sufficient for the daily ration of a four months old pig, and should make 1 lb. of L.W. increase. About 550 lb. skim milk equals 100 lb. meal.

Seedy-cut is a dark discoloration of the milk ducts in the belly, caused by the colouring matter in dark-skinned pigs.

Watch the live-stock, meat, grain and feed markets; not to do so frequently means an unfavourable balance-sheet.

Read the weekly stock papers. This keeps one abreast of the times by following the trend of what is happening in the pig world. There is always something to learn therefrom, and knowledge is capital.

CHAPTER XXI

THE OUTLOOK

UNFORTUNATELY there is much too great a discrepancy between what the farmer is paid for his pigs and what the public have to pay for the products thereof. I have in a previous chapter pointed out that the Danish farmers, who, chiefly by their export of bacon and ham meet our requirements, discovered this several years ago, and found the only solution in organizing and owning their own bacon factories, or, in other words, by co-operation. English farmers, however, for reasons somewhat obscure, appear to be shy of this method, but I believe it is only a matter of a short time when from sheer necessity they will be forced to avail themselves of this means of defence. The trusts, combines and rings have to be baulked of their prey.

Most of us know the tale of a visitor to a lunatic asylum, who on asking a warder if he were not afraid of the patients attacking him, replied, "Lunatics never organize." Of course, I am not suggesting, or even inferring, that farmers are lunatics; far from it, for they are the sanest class of people on earth, but I do maintain that it is not common sense to allow a very considerable profit to be filched from us when there is a tried and successful weapon of prevention at disposal.

Probably the £2,537,681 spent by this country on overseas frozen, salted, and tinned pork represents an average of ten thousand pigs per week. This, with the millions spent on bacon from Denmark, the United States and elsewhere, is sufficient proof that the pig industry in England has an unlimited home market, and one from which the foreigner anyhow should be, to a considerable extent, ousted.

Unfortunately, if outside competition were curtailed, under present conditions farmers would not reap the full benefit. Years ago they did obtain a price nearer to that of the retailers, but to-day the distributors are so well organized that they are able to set the lowest price on farm products when marketed and to exact the highest from the consumer.

Unless, therefore, there is a big change in the present selling system many farmers must sooner or later face financial ruin. A new system must inevitably come into vogue if English farming is to be carried on profitably, and without a doubt this should be co-operation, so far as pigs are concerned.

If producers are guided by the experience of those who raise pigs at the lowest cost of production by employing modern methods and supplying what is wanted, then together with the Danish system of co-operation one is induced to take an optimistic outlook for modern pig-keeping in this country.

PIG POPULATION AND OUTBREAKS OF SWINE FEVER
FOR THE YEARS 1926 AND 1929.

England, Wales & Scotland.	No. of pigs.	Out-breaks.	Pigs slaughtered on be- half of the Ministry.
1926	2,798,576	1,643	593
1929	2,508,040	2,981	3,677
1934	3,518,600	1,414	—
	(on June 4)		

Under the present policy of the Ministry, swine are only slaughtered when it is considered necessary or advisable to slaughter for the purpose of establishing the fact that swine fever does or does not exist on premises.

IMPORTED BACON, HAMS, PORK AND LARD INTO THE
UNITED KINGDOM IN THE YEAR 1933

	Quantity Cwts.	Value £
BACON		
From Lithuania	415,526	1,201,681
„ Sweden	402,634	1,332,157
„ Denmark	5,524,497	19,123,933
„ Poland	783,758	2,293,116
„ Netherlands	871,950	2,678,234
„ U.S.A.	62,931	170,964
„ Irish Free State	204,303	612,552
„ Canada	506,113	1,599,718
„ Other Countries	313,221	922,152
Total	9,084,933	£29,934,507
HAMS		
From Poland	74,056	246,219
„ U.S.A.	564,048	2,004,497
„ Argentine Republic	29,824	89,286
„ Irish Free State	20,206	64,790
„ Canada	180,639	657,842
„ Other Countries	711	4,574
Total	869,483	£3,067,208

THE OUTLOOK

IMPORTED BACON, HAMS, PORK AND LARD INTO THE
UNITED KINGDOM IN THE YEAR 1933

	Quantity Cwts.		Value £
PORK, FRESH			
From Irish Free State	194,695	..	498,634
„ Other Countries	——	..	——
Total ..	194,695	..	£498,634
PORK, FROZEN			
From U.S.A. ..	83,322	..	257,563
„ Argentine Republic	166,356	..	417,046
„ New Zealand ..	278,082	..	662,400
„ Other Countries	95,369	..	242,794
Total ..	623,129	..	£1,579,803
PORK, SALTED	21,003	..	£41,704
PORK, TINNED	174,856	..	£947,269
LARD			
From U.S.A.	2,501,548	..	4,301,385
„ Other Countries	379,898	..	659,234
Total ..	2,881,446	..	£4,960,619

The average price *per cwt.* of these imports for the past two years is as follows:

				1932				1933		
				£	s.	d.		£	s.	d.
Bacon	..	..	..	2	13	0	..	3	5	10
Hams	..	..	..	3	7	11	..	3	10	6
Pork, Fresh	..	..	..	2	15	7	..	2	11	3
,, Frozen	..	..	..	2	14	7	..	2	10	7
,, Salted	..	..	..	1	9	4	..	1	19	8
,, Tinned	..	..	..	5	17	4	..	5	8	4
Lard	..	..	..	1	19	8	..	1	14	5

The total weight and value of these imported pig products was 692,477 tons and £41,029,744 respectively. For 1932 the weight was 770,638 tons with a value of £40,318,817; 1933, therefore, shows a *decrease* of 78,161 tons, but an increase in value of £710,927.

Hams, frozen and tinned pork and lard had increases, but the bulk of the decrease was due to less bacon from Denmark. Frozen pork makes a big increase, New Zealand being the chief country of origin.

Lard still continues to come over in an enormous quantity and to satisfy this country's requirement 144,022 tons have to be imported at a cost of about £5,000,000 per annum! Apart from the Irish Free State, it is noticeable that of the total imports only some 48,000 tons and valued at £2,919,960 were of Empire production.

Annual amounts and values of Imported Pig Products since this book was first published, in 1924.

		Tons		Value £
1924	..	651,982	..	58,594,846
1925	..	628,297	..	66,588,566
1926	..	593,116	..	63,973,039
1927	..	621,207	..	54,186,211
1928	..	651,784	..	56,097,689
1929	..	636,522	..	60,877,235
1930	..	678,733	..	57,188,902
1931	..	779,934	..	46,091,755
1932	..	770,638	..	40,318,817
1933	..	692,477	..	41,029,744

Taking the killing weight of a pig at 170 lb. these annual amounts represent roughly 8,000,000 pigs, and to meet this country's demand the product of about 20,000 pigs must be imported on an average *every day.*

PIG SOCIETIES

Cumberland Pig Breeders' Association.— Sec., GEORGE M. BELL, 38 Lowther Street, Carlisle.

Essex Pig Society.—Mrs. URQUHART, 11 Galleywood Road, Chelmsford.

Gloucester Old Spots Pig Society.—Sec., A. E. PERKINS, 36 Baldwin Street, Bristol.

Large Black Pig Society.—Sec., B. J. ROCHE, 12 Hanover Square, London, W.1.

Large White Ulster Pig Society.—Hon. Sec., Balmoral House, Belfast.

National Lop Eared Pig Society.—Sec., S. A. YEO, 3 West Street, Okehampton, Devon.

National Pig Breeders' Association (Berkshire, Large White, Middle White, Tamworth and Wessex Saddleback Breeds),—Sec., ALEC HOBSON, 92 Gower Street, London, W.C.1.

Royal Dublin Society's Register of Pigs.— The Secretary, Balls Bridge, Dublin.

Welsh National Pig Society.—Sec., Capt. T. A. HOWSON, Queen Buildings, Queen Street, Wrexham.

BREEDERS' TABLE—115 DAYS

Date		When born		Date		When born	
January	1	April	25	February	1	May	26
,,	2	,,	26	,,	2	,,	27
,,	3	,,	27	,,	3	,,	28
,,	4	,,	28	,,	4	,,	29
,,	5	,,	29	,,	5	,,	30
,,	6	,,	30	,,	6	,,	31
,,	7	May	1	,,	7	June	1
,,	8	,,	2	,,	8	,,	2
,,	9	,,	3	,,	9	,,	3
,,	10	,,	4	,,	10	,,	4
,,	11	,,	5	,,	11	,,	5
,,	12	,,	6	,,	12	,,	6
,,	13	,,	7	,,	13	,,	7
,,	14	,,	8	,,	14	,,	8
,,	15	,,	9	,,	15	,,	9
,,	16	,,	10	,,	16	,,	10
,,	17	,,	11	,,	17	,,	11
,,	18	,,	12	,,	18	,,	12
,,	19	,,	13	,,	19	,,	13
,,	20	,,	14	,,	20	,,	14
,,	21	,,	15	,,	21	,,	15
,,	22	,,	16	,,	22	,,	16
,,	23	,,	17	,,	23	,,	17
,,	24	,,	18	,,	24	,,	18
,,	25	,,	19	,,	25	,,	19
,,	26	,,	20	,,	26	,,	20
,,	27	,,	21	,,	27	,,	21
,,	28	,,	22	,,	28	,,	22
,,	29	,,	23				
,,	30	,,	24				
,,	31	,,	25				

BREEDERS' TABLE—115 DAYS (*continued*)

Date		When born		Date		When born	
March	1	June	23	*April*	1	July	24
,,	2	,,	24	,,	2	,,	25
,,	3	,,	25	,,	3	,,	26
,,	4	,,	26	,,	4	,,	27
,,	5	,,	27	,,	5	,,	28
,,	6	,,	28	,,	6	,,	29
,,	7	,,	29	,,	7	,,	30
,,	8	,,	30	,,	8	,,	31
,,	9	July	1	,,	9	August	1
,,	10	,,	2	,,	10	,,	2
,,	11	,,	3	,,	11	,,	3
,,	12	,,	4	,,	12	,,	4
,,	13	,,	5	,,	13	,,	5
,,	14	,,	6	,,	14	,,	6
,,	15	,,	7	,,	15	,,	7
,,	16	,,	8	,,	16	,,	8
,,	17	,,	9	,,	17	,,	9
,,	18	,,	10	,,	18	,,	10
,,	19	,,	11	,,	19	,,	11
,,	20	,,	12	,,	20	,,	12
,,	21	,,	13	,,	21	,,	13
,,	22	,,	14	,,	22	,,	14
,,	23	,,	15	,,	23	,,	15
,,	24	,,	16	,,	24	,,	16
,,	25	,,	17	,,	25	,,	17
,,	26	,,	18	,,	26	,,	18
,,	27	,,	19	,,	27	,,	19
,,	28	,,	20	,,	28	,,	20
,,	29	,,	21	,,	29	,,	21
,,	30	,,	22	,,	30	,,	22
,,	31	,,	23				

BREEDERS' TABLE—115 DAYS (continued)

Date		When born		Date		When born	
May	1	August	23	June	1	Sept.	23
,,	2	,,	24	,,	2	,,	24
,,	3	,,	25	,,	3	,,	25
,,	4	,,	26	,,	4	,,	26
,,	5	,,	27	,,	5	,,	27
,,	6	,,	28	,,	6	,,	28
,,	7	,,	29	,,	7	,,	29
,,	8	,,	30	,,	8	,,	30
,,	9	,,	31	,,	9	October	1
,,	10	Sept.	1	,,	10	,,	2
,,	11	,,	2	,,	11	,,	3
,,	12	,,	3	,,	12	,,	4
,,	13	,,	4	,,	13	,,	5
,,	14	,,	5	,,	14	,,	6
,,	15	,,	6	,,	15	,,	7
,,	16	,,	7	,,	16	,,	8
,,	17	,,	8	,,	17	,,	9
,,	18	,,	9	,,	18	,,	10
,,	19	,,	10	,,	19	,,	11
,,	20	,,	11	,,	20	,,	12
,,	21	,,	12	,,	21	,,	13
,,	22	,,	13	,,	22	,,	14
,,	23	,,	14	,,	23	,,	15
,,	24	,,	15	,,	24	,,	16
,,	25	,,	16	,,	25	,,	17
,,	26	,,	17	,,	26	,,	18
,,	27	,,	18	,,	27	,,	19
,,	28	,,	19	,,	28	,,	20
,,	29	,,	20	,,	29	,,	21
,,	30	,,	21	,,	30	,,	22
,,	31	,,	22				

BREEDERS' TABLE—115 DAYS (continued)

Date		When born		Date		When born	
July	1	Oct.	23	*August*	1	Nov.	23
,,	2	,,	24	,,	2	,,	24
,,	3	,,	25	,,	3	,,	25
,,	4	,,	26	,,	4	,,	26
,,	5	,,	27	,,	5	,,	27
,,	6	,,	28	,,	6	,,	28
,,	7	,,	29	,,	7	,,	29
,,	8	,,	30	,,	8	,,	30
,,	9	,,	31	,,	9	Dec.	1
,,	10	Nov.	1	,,	10	,,	2
,,	11	,,	2	,,	11	,,	3
,,	12	,,	3	,,	12	,,	4
,,	13	,,	4	,,	13	,,	5
,,	14	,,	5	,,	14	,,	6
,,	15	,,	6	,,	15	,,	7
,,	16	,,	7	,,	16	,,	8
,,	17	,,	8	,,	17	,,	9
,,	18	,,	9	,,	18	,,	10
,,	19	,,	10	,,	19	,,	11
,,	20	,,	11	,,	20	,,	12
,,	21	,,	12	,,	21	,,	13
,,	22	,,	13	,,	22	,,	14
,,	23	,,	14	,,	23	,,	15
,,	24	,,	15	,,	24	,,	16
,,	25	,,	16	,,	25	,,	17
,,	26	,,	17	,,	26	,,	18
,,	27	,,	18	,,	27	,,	19
,,	28	,,	19	,,	28	,,	20
,,	29	,,	20	,,	29	,,	21
,,	30	,,	21	,,	30	,,	22
,,	31	,,	22	,,	31	,,	23

BREEDERS' TABLE—115 DAYS (continued)

Date		When born		Date		When born	
Sept.	1	Dec.	24	October	1	January	23
,,	2	,,	25	,,	2	,,	24
,,	3	,,	26	,,	3	,,	25
,,	4	,,	27	,,	4	,,	26
,,	5	,,	28	,,	5	,,	27
,,	6	,,	29	,,	6	,,	28
,,	7	,,	30	,,	7	,,	29
,,	8	,,	31	,,	8	,,	30
,,	9	Jan.	1	,,	9	,,	31
,,	10	,,	2	,,	10	Feb.	1
,,	11	,,	3	,,	11	,,	2
,,	12	,,	4	,,	12	,,	3
,,	13	,,	5	,,	13	,,	4
,,	14	,,	6	,,	14	,,	5
,,	15	,,	7	,,	15	,,	6
,,	16	,,	8	,,	16	,,	7
,,	17	,,	9	,,	17	,,	8
,,	18	,,	10	,,	18	,,	9
,,	19	,,	11	,,	19	,,	10
,,	20	,,	12	,,	20	,,	11
,,	21	,,	13	,,	21	,,	12
,,	22	,,	14	,,	22	,,	13
,,	23	,,	15	,,	23	,,	14
,,	24	,,	16	,,	24	,,	15
,,	25	,,	17	,,	25	,,	16
,,	26	,,	18	,,	26	,,	17
,,	27	,,	19	,,	27	,,	18
,,	28	,,	20	,,	28	,,	19
,,	29	,,	21	,,	29	,,	20
,,	30	,,	22	,,	30	,,	21
				,,	31	,,	22

BREEDERS' TABLE—115 DAYS (*continued*)

Date		When born		Date		When born	
Nov.	1	Feb.	23	Dec.	1	March	25
,,	2	,,	24	,,	2	,,	26
,,	3	,,	25	,,	3	,,	27
,,	4	,,	26	,,	4	,,	28
,,	5	,,	27	,,	5	,,	29
,,	6	,,	28	,,	6	,,	30
,,	7	March	1	,,	7	,,	31
,,	8	,,	2	,,	8	April	1
,,	9	,,	3	,,	9	,,	2
,,	10	,,	4	,,	10	,,	3
,,	11	,,	5	,,	11	,,	4
,,	12	,,	6	,,	12	,,	5
,,	13	,,	7	,,	13	,,	6
,,	14	,,	8	,,	14	,,	7
,,	15	,,	9	,,	15	,,	8
,,	16	,,	10	,,	16	,,	9
,,	17	,,	11	,,	17	,,	10
,,	18	,,	12	,,	18	,,	11
,,	19	,,	13	,,	19	,,	12
,,	20	,,	14	,,	20	,,	13
,,	21	,,	15	,,	21	,,	14
,,	22	,,	16	,,	22	,,	15
,,	23	,,	17	,,	23	,,	16
,,	24	,,	18	,,	24	,,	17
,,	25	,,	19	,,	25	,,	18
,,	26	,,	20	,,	26	,,	19
,,	27	,,	21	,,	27	,,	20
,,	28	,,	22	,,	28	,,	21
,,	29	,,	23	,,	29	,,	22
,,	30	,,	24	,,	30	,,	23
				,,	31	,,	24

MEMORANDA

Sow's Name or Number	Sire	When Bred	Due to Farrow	Number of Litter	Sex

MEMORANDA

Sow's Name or Number	Sire	When Bred	Due to Farrow	Number of Litter	Sex

INDEX

INDEX

Lightning Source UK Ltd.
Milton Keynes UK
18 August 2009
142808UK00001B/130/A